Alim-un- Nisa
Muhammad Muneeb Bilal
Fatima Sajjad

Fantasies of Licorice

Alim-un- Nisa
Muhammad Muneeb Bilal
Fatima Sajjad

Fantasies of Licorice

Noor Publishing

Imprint
Any brand names and product names mentioned in this book are subject to trademark, brand or patent protection and are trademarks or registered trademarks of their respective holders. The use of brand names, product names, common names, trade names, product descriptions etc. even without a particular marking in this work is in no way to be construed to mean that such names may be regarded as unrestricted in respect of trademark and brand protection legislation and could thus be used by anyone.

Cover image: www.ingimage.com

Publisher:
Noor Publishing
is a trademark of
Dodo Books Indian Ocean Ltd. and OmniScriptum S.R.L Publishing group
Str. Armeneasca 28/1, office 1, Chisinau-2012, Republic of Moldova, Europe
Printed at: see last page
ISBN: 978-620-4-72344-0

ARTICLE

ON

FANTASIES OF LICORICE

Authors

Ms. Alim-un-Nisa

(Principal Scientific Officer)

Muhammad Muneeb Bilal

Fatima Sajjad

Contents:

Introduction:

The flowering plant Glycyrrhiza glabra, which belongs to the Fabaceae family of beans, is known by its common name, licorice. Its root can be used to make a flavoring that is sweet and aromatic.

Western Asia, North Africa, and Southern Europe are the native habitats of the perennial herbaceous legume known as licorice. It is not botanically connected to fennel or anise, which are also producers of flavouring chemicals. In various West Asian and European nations, licorice is used as a flavouring in candies and tobacco.

Extracts of licorice have been utilised in traditional medicine and herbalism. Licorice overconsumption should be clinically recognised in individuals who come with hypokalemia and muscle weakness that are otherwise unexplained (more than 2 mg/kg/day of pure glycyrrhizinic acid, a licorice component). Excessive licorice eating has been linked to mortality in at least one instance.

The German commission has approved an extract of Glycyrrhiza glabra roots, which is frequently used in traditional Siddha therapy today. Gastritis, coughing, pneumonia, ulcers, inflammation, and epilepsy can all be treated with licorice. Licorice use and hypertension are known to be associated. It is a result of hypokalemia, edoema, and sodium retention.

Figure 1: licorice root

History:

Origin of licorice is trace back to 2300 BC. Licorice was described as a magical herb that could invigorate ageing men in the Divine Farmer's Herb-Root Classic, Pen-Ts'ao, which was written by China's Emperor Shennong.

In addition to other valuables, licorice root was also discovered in the tomb of the Egyptian pharaoh Tutankhamun, who lived in 1350 BC. Its actual reason is still a secret. We do, however, know that a common component of many of the ancient Greek cures was licorice. Cough medication is one example. Due to its pain-relieving and anti-inflammatory properties, licorice was to perform this role in modern society, where it became a common ingredient in cough syrup.

In 1872, Galle & Jessen and in 1902, Hgh's licorice factory were established. Both companies began by making boiled candies before expanding into the production of licorice and jellies. From a cough remedy to confectionery, licorice has a long history in Italy, France, Germany, and England. There are numerous examples of licorice's evolution during the 19th century that serve as reminders of its contemporary use. One of the earliest licorice products still in existence is Ga-Jol from 1933. It was introduced as a pastille for the throat to treat coughing and hoarseness. In the beginning, soft licorice rods were more popular as children's candy and hard licorice pastilles like Ga-Jol were meant to soothe sore throats.

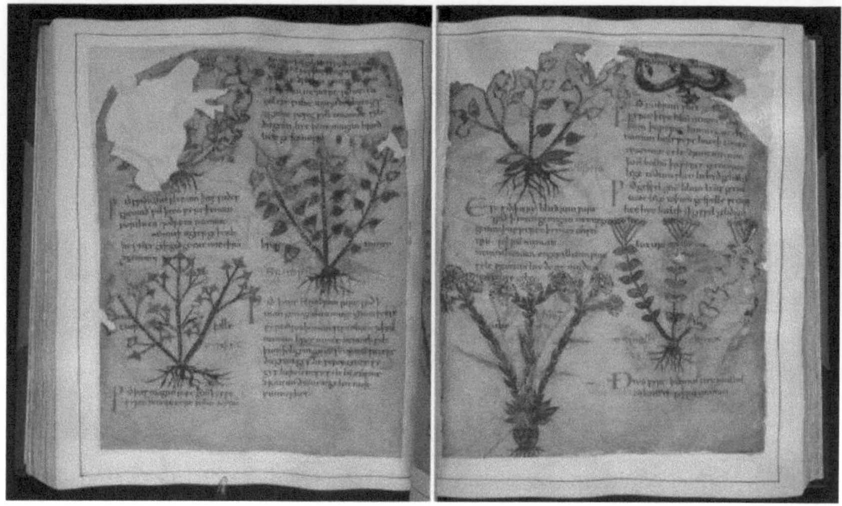

Figure 2: thousand year old remedy

Geographical distribution:

The Mediterranean region, central and south-western Asia, and Eurasia are the native habitats of Glycyrrhiza glabra. According to Hayashi and Hayashi and Sudo, G. glabra is found in South Europe (Spain, Italy), Turkey, Iran, Iraq, Central Asia, and the north-western portion of China, whereas G. uralensis is only found in the north-eastern section of the country, in the Xinjiang Uygur Autonomous Region. The two varieties of G. glabra are G. glabra var. glandulifera (Russian licorice) and G. glabra var. typica (Spanish licorice). There are three known varieties of G. glabra; the Spanish and Italian licorices are classified as G. glabra var. typica, while the Russian licorice is classified as G.

Glabra variety glandulifera; according to G., Turkish and Persian licorice Glabra variety violacea. Iran, Afghanistan, Pakistan, Iraq, Azerbaijan, Uzbekistan, Turkmenistan, and the People's Republic of China are among the nations that produce licorice. The three aforementioned species are used to make commercial licorice in China.

Figure 3: geographical distribution of licorice

Taxonomy hierarchy:

1. Domain: Eukaryota

Due of its membrane-bound cells that include nuclei, the licorice plant falls under this category. Eukaryotic organisms have a number of organelles that are membrane-bound, including mitochondria and chloroplasts.

2. Kingdom: Plantae

Because it employs photosynthesis to produce energy and because it contains chlorophyll in its cells, the licorice plant belongs to this kingdom. Cell walls are also present in members of this kingdom, which contribute in the support of the plant.

3. Phylum: Tracheophyta

The licorice belongs to this phylum since it is a vascular plant with vascular tissue that carries nutrients and water throughout the entire plant. Vascular plants can develop higher and are more effective at using and gathering resources.

4. Class: Magnoliopsida

This class of organisms also includes dicots, and they share several characteristics with other dicots. The stems of plants in this class typically have a lot of vascular cambium and are woody. This type of plants often has a primary root system and leaves with net veins. Also, pollen grains often have three holes, and floral organs frequently appear in multiples of five.

5. Sequence: Fabales

The order's members are classified as dicots with complex leaves. Although it's not always the case, plants in this group typically produce blooms with a single carpel that develops into a legume. To fix nitrogen, plants in this group frequently develop symbiotic interactions with microbes.

6. Family: Leguminosae

This family includes the licorice plant because of the type of fruit it produces. The fruit is a dried fruit, often known as a pod. Additionally, these plants have the capacity to fix nitrogen thanks to a symbiotic interaction with specific bacteria.

7. Genus: Glycyrrhiza

All of the plants in this genus belong to the Legume family. They can all be found in Asia, Australia, Europe, and the Americas, and they all produce legumes.

8. Specie: Glycyrrhiza glabra

The licorice plant is the name of this species. It is well-known for producing glycyrrhizic acid, a substance that is present in the plant's roots. This substance is a naturally occurring sweetener that is 30–50% sweeter than sucrose.

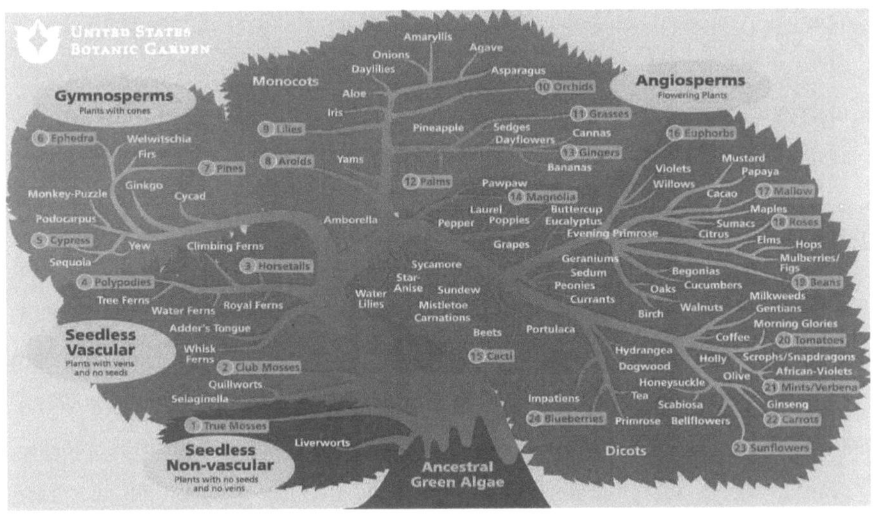

Figure 4: phylogenetic tree of licorice

Phytochemistry of licorice:

In terms of nutrition, licorice is a good source of proteins, amino acids, simple sugars, polysaccharides, and mineral salts (including calcium, phosphorus, sodium, potassium, iron, magnesium, silicon, selenium, manganese, zinc, and copper). It also contains pectins, resins, starches, sterols, and gums. There have been reports of oestrogens, tannins, phytosterols (sitosterol and stigmasterol), coumarin, vitamins (B1, B2, B3, B5, E, and C), and glycosides. Numerous biological substances have also been discovered, primarily triterpenes, saponins (which give substances their sweet flavour), and flavonoids. While the aglycones are present as oleananes, the licorice saponins are present as glucuronides. The primary distinguishing elements of licorice are its triterpene saponins, which are responsible for its sweet flavour.

The chief active component of roots is glycyrrhizin, a triterpenoid saponin that is almost 50 times sweeter than sucrose. Glycyrrhizin, a combination of 2% to 25% potassium, calcium, and magnesium salts of glycyrrhizic acid, makes up around 10% of the dry weight of licorice root. Intestinal bacteria convert glycyrrhizin after oral dose into 18-glycyrrhetic acid 3-omonoglucuronide and glycyrrhetic acid.

Licorice's yellow colour is a result of its flavonoid concentration. The detected flavonoids fall into a variety of categories, including flavanones, flavones, flavanonols, chalcones, isoflavans, isoflavenes, isoflavones, and isoflavanones. The main flavonoids are liquiritin, isoliquiritin, and glycosides of liquiritigenin (4′, 7′-dihydroxyflavanone) and isoliquiritigenin (2′, 4, 4′-trihydroxychalcone), respectively. Prenyllicoflavone A, shinpterocarpin, glucoliquiritin apioside, shinflavanone, and 1methoxyphaseolin are the five novel flavonoids that have been extracted from dried roots. Additionally, the leaves' pinocembrin and licoflavanone were isolated. The main isoflavone discovered, glabridin, makes from between 0.08% and 0.35% of the dry weight of roots. Isoprenoid-substituted flavonoids, chromenes, coumarins, dihydrostilbenes, coumestans, benzofurans, and dihydrophenanthrenes are the minor phenolic chemicals. The unique odour of roots is further enhanced by the presence of numerous volatile substances as geraniol, pentanol, hexanol, terpinen-4-ol, and terpineol. Additionally abundant in propionic acid, benzoic acid, furfur aldehyde, 2, 3-butanediol, furfuryl formate, maltol, 1-methyl-2-formylpyrrole, and trimethylpyrazine is G. glabra essential oil.

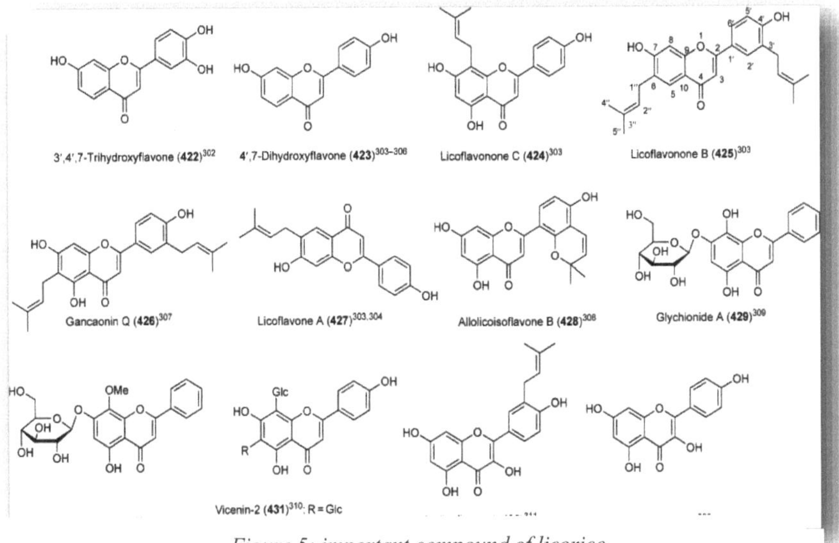

Figure 5: important compound of licorice

Types:

1. Black licorice:

Candy with an anise flavour and a black or dark tint is referred to as "black licorice" in both Canada and the US. These sweets can be flavoured with extracts of licorice, anise, or both. Anise extract, as opposed to licorice root, is most frequently used to flavour black licorice created in the United States.Traditional licorice root extract can be found in several European black licorice candy variations, including those from Iceland, Finland, and the Netherlands. Because all licorice is black or dark in colour in these nations, the word "black" is redundant. In fact, to distinguish it from salty licorice, what is referred to as black licorice in North America is known as sweet licorice in Scandinavia, the Netherlands, and northern Germany.

What ingredients make licorice? Glycyrrhizin, the substance that gives licorice root its sweet flavour, is a need for licorice to be considered really traditional. Traditional black licorice also contains sugar and a binder, which is often some sort of starch or flour, though occasionally gum arabic or gelatin is used in place of or in combination with flour. In some versions, beeswax is added to provide a glossier candy surface, and molasses is occasionally added to enhance the licorice flavour.

Figure 6 Black licorice

2. Red licorice:

Red licorice is a chewy candy that's prepared using a method similar to that of conventional licorice but without the addition of licorice root or anise.

Black and red licorice both come in a majority of the same shapes. Traditional licorice and red licorice sweets differ primarily in colour and flavourings, which are typically strawberry but can also occasionally be cherry, raspberry, or cinnamon. Many products using red licorice use both natural and artificial flavourings.

Figure 7: red licorice

3. Salted licorice:

Known as salmiaklakrits or saltlakrits in Swedish or salmiaklakrids or saltlakrids in Dutch, salty licorice. This sweet is frequently referred to as salzlakritz in German and salmiakki in Finnish. In any of these nations, you might just just call it "salmiak," and people would understand what you meant.With one difference, the components used to make this Dutch and Scandinavian licorice are the same as those used to make conventional sweet licorice and are flavoured with licorice root. Ammonium chloride, often known as sal ammoniac, is used to give salty licorice its salty flavour.

The amount of ammonium chloride in the confection—up to 8% depending on the brand and the country of origin—determines the strength of both the saltiness and the accompanying astringency. salmiak salt or powder, and occasionally table salt, sodium chloride, are sprinkled on top of extra salty licorice.Typically black or very dark brown in colour, salty licorice sweets come in a range of forms and textures. In addition, it is employed to flavour a wide range of additional delicacies, including ice cream, chewing gum, and even alcoholic beverages.

But at Licorice.com, we have a wide selection of salted licorice for every taste. Try our Honey Bees if you've never tried salty licorice before. This candy is honey-sweetened and has a tiny salty kick towards the end. Try our Farm Salted Assortment for a saltiness that is more pronounced, and if you truly enjoy salty licorice (or want to challenge your palate), try our Salties with dubble zout or double salt.

Figure 8: salted licorice

Cultivation and growth:

In several regions of Baluchistan, Chitral, and the Hindu Kush Himalayan regions, licorice grows naturally. It Chitral's roots are substantial and of comparable quality to the best licorice for commercial use is Chinese licorice. The systematic production and collecting of medicinal plants is not given any attention in Pakistan, and the only reliable data that are available are a few papers in the Pakistan Journal of Forestry. Although many suggestions have been made at the institutional level and even supported through research and development in this regard, there is no national overview of the species that should be cultivated in the country. At Pakistan Forest Institute (PFI), cultivation studies of numerous medicinal plants were conducted. The yield of roots of G. glabra was 4200 lbs/acre, while the active principle glycyrrhizin was 1.2% with 10% blood sugar.

The province of Baluchistan provides a highly favourable environment for its cultivation. If the market structure for medical plants is developed and cultivation of medicinal plants is carried out according to scientific principles, medicinal plants can offer higher income to the local people than the conventional crops of the areas. Other plants, like Foeniculum vulgare, were also farmed, although not extensively. This plant needs to be grown in the Chitral valleys of Arkari and Golengol. For good production, it needs a deep, well-cultivated, fertile, moisture-retentive soil. Plants are resilient down to roughly -15oC. It did not thrive in clay and preferred a sandy soil with lots of moisture. The herb flourished in the maritime environment. The edible root of licorice, which is also widely used in flavouring and medicine, is frequently grown.

In Russia, Glycyrrhiza glandulifera can sprout adventitious roots that are up to 10 cm thick. This species coexists in symbiosis with specific soil bacteria, which cause nodules to develop on the roots and fix atmospheric nitrogen. While the growing plant uses some of this nitrogen, other neighbouring plants can also use some of it. In Mastung, licorice was planted in irrigated and rainfed sand dune areas between 1981 and 1982, according to Mohammad and Rehman (1985). After one year, the survival rates for these two sites were 92 and 85%, respectively. In the first two years, rhizome and root growth greatly increased. The study came to the conclusion that this plant might prove to be an appropriate forage plant for stabilising sand dunes in the Mastung area. Rhizomes and roots are often removed from plants around October, ideally from those that have not yet produced any fruit. The medication is cleaned after the buds and rootlets are removed. Peeled and cut into smaller pieces are some of the parts. The medication is dried twice, once in the sun and once in the shade, losing around half its weight each time.

- **How to Grow Your Own Licorice:**

  At the end of February, you can sow licorice seeds indoors early. Just keep in mind to chill the seeds for two to four weeks prior to steeping them in water for 24 hours to boost germination.

  After that, spread a thin layer of dirt over the seeds in the container. A loose soil with little nutrients, such as our Plantura Organic Herb & Seedling Compost, is perfect at this time. Any additional nutrients the compost provides will not be able to be utilised by the seeds. Keep the soil in your container damp but not soggy, and place it in a warm (20°C) area. The seeds will start to sprout in 15 to 30 days. It is important to give licorice plants plenty of space to grow when they are planted outside at the end of May. Their root systems and rhizomes are strong and extensive. In our experience, it is best to give each seedling 50x50cm.

In general, it is possible to grow licorice plants in pots. However, the pots must be large enough to allow the expansive taproots to grow. This is especially important if you would like to harvest licorice roots. However, if you plan to keep licorice indoors as and ornamental instead, install rhizome barriers to slow the plant's growth.

Figure 9: licorice growth

- **Licorice plant care:**

Licorice is simple to maintain. It will only require fertilising once a year, in the spring, if you plant it in soil that is rich in nutrients. We advise using compost or a slow-acting organic fertiliser like our Plantura All Purpose Plant Food. Importantly, stay away from general herb fertilisers since they frequently have insufficient phosphorus.

Make sure to plant your licorice plant in well-drained soil when you are growing it at home. The licorice plant dislikes being submerged in excess water. If that happens, your plant can begin to degrade. Once planted, your plant can withstand a drought, but it loves routine watering.

This plant is a perennial, so unless you live in a tropical climate and can overwinter it, it won't flourish. When licorice bloom it flower appears to be tiny and white in appearance. In more tropical areas of the U.S., employ caution as this plant has the potential to self-seed and become slightly invasive. Remove the blossoms if that's a problem in your area because they aren't really ornamental.

The semi-trailing or cascading growth pattern of this plant makes it ideal for hanging baskets and containers. Check a plant's habit before buying if you plan to use it as a spiller because certain types have a more upright habit. Licorice plants respond well to pruning, so it's a good idea to pinch trailing kinds when they're young to promote healthy branching.

Figure 10: care of licorice

- **Harvesting:**

  Once your plant is two years old, you can begin to harvest the licorice root. This is best done in the fall after the growth season has ended. With a sharp spade, dig up the plant, and using clean secateurs or scissors, trim part of the roots. Before using the roots to produce a delightful, healthy tea, let them dry for a few months. Be careful not to injure the major roots that are connected to the developing plant. By doing this, you can replant it in the garden and enjoy the benefits again in a few years.

Drying licorice:

1. Air drying:

   Herbs are frequently dried via air drying because it allows air to circulate around them, preventing mould growth. Licorice roots should be gathered in a compact group and tied together with string. To properly dry them, hang them up for three to six weeks in a cool, dry location.

2. Sun drying:

   In addition to air drying, licorice root can also be dried in the sun, which is a little faster. Use a pair of clean, sharp garden shears or scissors to cut the roots into little pieces that are around an inch (2.5 cm) long. Put the pieces of root on a tray and place it in a wind- and sun-protected area. In around two weeks, the roots ought to have dried up.

3. Dehydration:

   The roots can be dried the quickest by using a dehydrator, but you must pay strict attention to them. Make sure the cut roots are evenly spaced out before placing them on the dehydrator trays. Put the dehydrator's lid on and choose a low temperature setting. After a few hours, check the roots and, when they are dry, pull them out. Dried licorice roots keep for up to a year in an airtight glass container.

Figure 11: harvesting

Medicinal uses:

- Canker sores:

A canker sore is a painful ulcer that develops on the gums or soft tissue of the mouth. These so-called "aphthous ulcers" are distinct from "cold sores" brought on by the herpes virus. Canker sores frequently fade away on their own without therapy, however many patients take lysine to hasten the healing process.

Try licorice for a canker sore if you think you have one. But don't simply purchase any licorice. Instead of those sugary pieces in the candy section, get licorice of higher quality for supplements. Those generally include anise, however. You should look for a chewable licorice supplement with the name DGL. Deglycyrrhizinated licorice is referred to as DGL. Deglycyrrhizinated refers to the removal of glycyrrhizin, a naturally occurring compound in licorice.

Licorice can apply a barrier over canker sores to stop saliva and other irritants from hindering the healing process. Simply indulge in a few licorice chews each day.

An investigation into the use of licorice root for the treatment of canker sores was conducted in 1989.Unfortunately, most patients could not tolerate the flavour or concentration of the licorice root extract in a warm water solution. It was simply too powerful. Unfortunately, despite the fact that licorice extract was a highly successful treatment for canker sores, it was challenging to persuade patients to use licorice consistently.

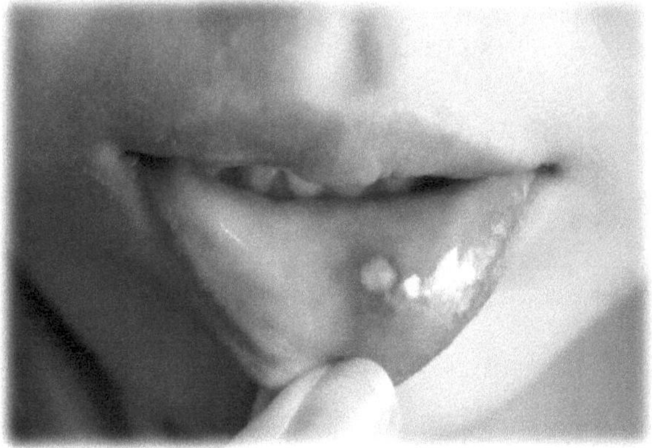

Figure 12: canker sores

- Chronic bronchitis:

An herb called licorice thrives in some regions of Asia and Europe. Glycyrrhizin, an ingredient in licorice root, can have negative effects if consumed in excessive quantities. Licorice is believed to contain compounds that reduce inflammation, alleviate coughing, and boost the body's natural ability to heal ulcers.

There is some data that suggests licorice root may help persons with chronic obstructive pulmonary disease reduce the progression of chronic bronchitis (COPD). The airways get inflamed over time as a result of chronic bronchitis.

Licorice root has been the subject of test tube research by researchers at Taiwan's Chung Shan Medical University. According to their findings, licorice root's glycyrrhizic, asiatic, and oleanolic acids have an antioxidant impact that may protect the cells in the bronchi that lead to the lungs.These findings might imply that, when taken in conjunction with conventional therapies, licorice can help reduce the progression of COPD. These findings need to be supported by additional human study.

Figure 12: Product for Chronic Bronchitis

- Colorectal cancer:

  Some researchers think that licorice's antioxidant properties may reduce the risk of some cancers, particularly colon cancer. While most studies have been conducted on animals or in test tubes, some findings have been encouraging. This includes a mouse study that revealed licorice root may have advantages in avoiding cancers linked to colitis.

  When administered to BALB/C mice that had been injected with CT-26 colon cancer cells, licorice extract greatly slowed the growth of the tumours. The therapeutic effectiveness of cisplatin was reduced when licorice extract and cisplatin were combined, but the anticancer activity of the licorice extract was significantly increased.

  A class of adaptable compounds known as licorice flavonoids affects cell growth, survival, and signalling in a variety of ways. Many flavonoids exhibit growth-inhibitory properties toward cancer cells, making them highly therapeutic in the treatment of cancer. To evaluate their in vivo effectiveness and potential toxicity, additional preclinical research are still required. It is also critical to assess how licorice flavonoids affect the metabolism of other medications and investigate any potential synergistic mechanisms.

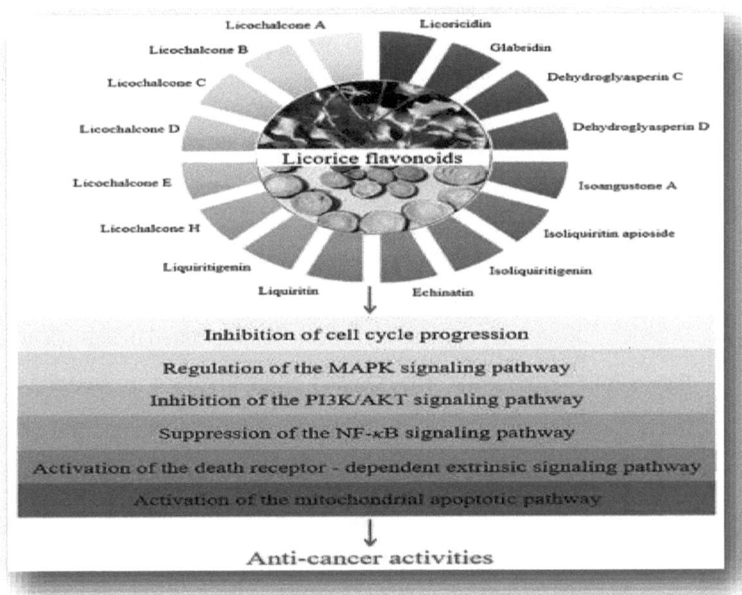

Figure 14: anticancer activity of licorice

- Functional Dyspepsia:

Licorice root may lessen the discomfort caused by functional dyspepsia (FD). This condition is characterized by episodes of discomfort in the upper abdomen.

In one trial, 50 participants in India received either a placebo (an inert "sugar pill") or a licorice root extract supplement twice day at a level of 75 milligrams (mg). Participants in the licorice-taking group reported more symptom improvement than those in the control group.

A perennial herb from the temperate zone known as G. glabra is said to have a number of pharmacological qualities, including demulcent, anti-inflammatory, and antiulcer actions. The ulcer index, volume, and total acidity of stomach contents all significantly improved in a preclinical research on GutGard in albino Wistar rats. Despite the well-known, beneficial effects of G. glabra on the digestive tract, there aren't enough clinical investigations on the effects of licorice and licorice preparations as a whole in individuals with functional dyspepsia. The Nepean dyspepsia score was considerably lower on days 15 and 30 after taking GutGard (75 mg) twice daily for 30 days compared to the placebo group.

Supplementing with GutGard has been demonstrated to significantly lower the overall symptom scores. Dietary supplements with licorice as one of the main constituents have also demonstrated notable effectiveness in treating patients with functional dyspepsia. The therapeutic reduction in gastrointestinal symptom-scores was also supported by Melzer et AL comprehensive's analysis of the effectiveness and tolerability of Iberogast. The number of patients with no symptoms or significant improvements, as determined by the patient assessments of overall efficacy, has demonstrated the superiority of the GutGard treatment (56%) over the placebo (0%) over functional dyspepsia. The enhancement of quality of life is considered a secondary outcome measure in existing clinical trials, but it might be the main goal of future research.

- Peptic Ulcers:

The scientific community is becoming more and more interested in the use of licorice to treat peptic ulcer disease. In particular, scientists were curious about how it affected the Helicobacter pylori bacterium (H. pylori). It is a challenging illness to treat and the main cause of peptic ulcers.

Licorice root, when combined with the common triple antibiotic therapy, completely eradicated H. pylori in 120 Iranian patients in a 2016 study. Only 62.5% of patients in a group receiving antibiotic medication and a placebo experienced successful treatment.

According to laboratory experiments, licorice root appears to have antibacterial properties. This indicates that it might be helpful in treating some bacterial infections like Staphylococcus aureus as well as other difficult-to-treat fungal infections like Candida albicans.

Licorice extract benefits for skin:

Licorice root is where black licorice candy comes from, but it's also the source of extract used topically for skin. This extract is filled with a variety of beneficial compounds that do everything from deliver antioxidant and anti-inflammatory effects to helping fade dark spots.

Reduces tyrosinase synthesis to prevent discoloration:

The formation of melanin, often known as pigment or colour, is a complex process, but at its core is an enzyme called tyrosinase. Tyrosinase production is inhibited by licorice extract, which also prevents the development of black spots. Licorice extract also lightens the skin in another way by removing excessive melanin. Liquiritin, an active ingredient that aids in dispersing and removing any existing melanin in the skin, is present, according to Chwalek. In other words, it can both assist diminish existing spots and aid in the prevention of new ones from appearing.

As a strong antioxidant:

Licorice includes a flavonoid, an antioxidant-rich component that reduces reactive oxygen species, which age and damage cells, like many other plant-based extracts.

Control oil production in skin:

Although this isn't one of the more widely acknowledged advantages, Chwalek claims that there is some evidence to support that licochalcone may help reduce oil production in the skin. An additional advantage of a substance would be that it controls oil output. It might even be the reason licorice extract is frequently used as a dandruff remedy in Ayurvedic medicine.

Figure 15: skin usage

Industrial uses:

The flavor of licorice is obtained from either the root or the extract. The extract's liquor is frequently processed into a powder or a solid block with a higher concentration. In the food sector, licorice root and extract are frequently used as sweeteners and flavorings in candies, gum, and beverages. Due to its therapeutic qualities, licorice is also utilized in both conventional and alternative treatments, such as cough syrups and over-the-counter medications.

Licorice is among the highly significant medicinal plant species. In 2007, the value of licorice trade was around 42 million US$, and evaluation of glycyrrhizin as a natural sweetener has added to its use. The cosmetics industry too is using the flavonoids isolated from licorice on a large scale.

- Pharmaceutical preparation:

The plant may be a source of new medications and therapeutic agents for the treatment of a number of diseases and afflictions, according to the different investigations on its bioactivities conducted by ethnobotanists, phytochemists, and experimental pharmacologists. This is a report of numerous actions.

Glycyrrhizin is a prescription medication used in Japan to treat liver and allergy conditions, according to Hayashi and Sudo . A Japanese business, Minophagen Pharmaceutical Co. Ltd., produces it as an injectable preparation (Stronger Neo-Minophagen® C) and in tablet form (Glycyron®). Since more than 60 years ago, stronger Neo-Minophagen® C has been offered in the Japanese market. Additionally, it is accessible in countries with a relatively high viral hepatitis B and C prevalence including China, Korea, Taiwan, Indonesia, India, and Mongolia. Licorice extracts, glycyrrhizin, and glycyrrhetinic acid are ingredients in a number of over-the-counter medications, such as antihistamines and inflammatory medications. Additionally, the glycyrrhetinic acid 3-O-hemisuccinate (carbenoxolone) prescription medication is used to treat peptic ulcer disease in England.

- Cosmetics:

In Japan, glycyrrhizin derivatives and licorice extracts are frequently employed in the creation of cosmetics. For their anti-inflammatory properties, glycyrrhizin, powdered glycyrrhiza roots, glycyrrhiza extracts, glycyrrhetinic acid, stearyl glycyrrhetinate, pyridoxine glycyrrhetinate, and glycyrrhetinic acid 3-O-hemisuccinate (carbenoxo Additionally, glabridin-containing glycyrrhiza flavonoids extracted from G. glabra are utilised in cosmetic products due to their abilities to lighten, calm, and reduce inflammation in the skin.

Licorice is a frequent and practical ingredient in the cosmetic industry. It functions as an emollient, a calming agent, and a depigmenting agent (Nomura et al. 2002). Jeanne Rose, a contemporary herbalist, suggests using licorice, comfrey, and chamomile or lavender in a steam facial. In a clinical investigation, topical application of a gel containing 2% licorice extract significantly reduced erythema, edoema, and pruritus (Saeedi et al. 2003). Additionally, its extracts are employed as skin lightening agents and are successful in treating post-inflammatory hyperpigmentation especially that brought on by chemical peeling and laser treatment (Callender et al. 2011). It aids in pore opening and facilitates skin penetration for other pore-cleansing and skin-healing herbs.

According to reports, using licorice root as an ingredient in shampoo can delay the appearance of an oily sheen by preventing the flow of scalp sebum for a week after shampooing. It is also a sweetening and flavouring ingredient in mouthwash and toothpaste. It is occasionally combined with anise and used in liqueurs and herbal teas.

The functional effects on scalp care for a scalp tonic containing licorice were documented in the review written by Shivakant and published in 2020. A common scalp issue called dandruff is linked to flaky and irritated skin. In a clinical trial, piroctone, olamine, and licochalcone A were used in combination by 102 patients (males, 56, and females, 46), all of whom had moderate to severe dandruff. A cytokine analysis was done in this study, and the results showed that using the evaluated treatments significantly reduced pro-inflammatory dandruff indicators. Additionally, the anti-fungal activity of test goods was discovered, showing a notable decrease in Malassezia colony-forming units following use of the anti-dandruff shampoo. The combined effect's advantage was mostly due to licochalcone's well-known anti-inflammatory properties.

- Anti-Acne properties:

A common skin condition known as acne vulgaris, which also affects the pilosebaceous unit, increases sebum production by sebaceous glands and results in aberrant hair follicle desquamation as a result of rising androgen levels that occur with the onset of puberty. Natural medicines frequently have less unpleasant side effects than manufactured ones.

G. glabra appears to be a unique acne treatment. G. glabra was one of the oriental herb extracts whose anti-acne therapeutic activities were examined in terms of their antichemotactic, antilipogenic, and antibacterial properties against Propionibacterium acnes on polymorphonuclear leucocytes. In contrast to the pronounced development of resistance in bacteria treated with erythromycin, G. glabra demonstrated exceptional antibacterial activity against P. acnes with low induction of resistance.

- Food additives:

Cyclodextrin glucanotransferase is used to create enzymatically modified licorice extract (EMLE), a natural sweetener. It is utilised because to its distinct qualities, which include greater taste and higher solubility than licorice extract. Six significant components that were extracted from EMLE had their structures elucidated and their sweetness was examined in the current work. Glycyrrhizin (1), 3-O-[beta-D-glucuronopyranosyl-(1—>2)-beta-D-glucuronopyranosyl) liquiritic acid (2), and their derivatives glucosylated at the C-4 position of the terminal glucuronopyranose with additional one (3 and 4, respectively) and two (5 and 6, respectively) glucose moieties were the isolated compounds. The two main and minor sweet components of licorice extract are compounds 1 and 2, respectively. The first novel compounds to be isolated are compounds 3-6. Compared to compound 1, compound 2, was sweeter. Contrary to common belief, compound 3, a monoglucosylated derivative of compound 1, was sweeter than compound 4. Both compounds were less sweet than their parent compounds, but the sweetness that lingered after swallowing was noticeably sweeter. The glucose moieties of compounds 5 and 6, which had two extra moieties, did not exhibit much sweetness.

Foods are given a sweet flavour by glycyrrhizin, which also has salt-softening, flavor-enhancing, and heat stability qualities. The long-lasting sweetness of glycyrrhizin is disliked by the majority of Japanese people, but it can be combined with natural sugars or other sweeteners to produce a sweetness that is more tolerable. As a result, glycyrrhizin and licorice extracts are utilised as food additives in a range of foods. Licorice extracts are found in sauces, instant noodles, and snack foods. To increase the sweetness of sweet foods like ice cream, sherbet, and sweet snacks, glycyrrhizin is added.

 In Japan, it is also used to lessen the saltiness of salty dishes such sausages, soy sauce, other sauces, savoury snacks, Kamaboko (boiled fish paste), Tsukudani (fish boiled in soy sauce), and tsukemono (Japanese pickles). Enzymatically hydrolyzed licorice extract (glycyrrhetinic acid 3-O-glucuronide) and enzymatically modified licorice extract (-glycosyl-glycyrrhizin) are also utilised as sweeteners in Japan. Because it has a higher solubility and better flavour than the untreated licorice extract, the former is created by processing the extract with cyclodextrin glycosyltransferase and is used as a sweetener. The latter is produced by hydrolyzing the licorice extract enzymatically. This licorice's sweetness is due to glycyrrhetinic acid 3-O-glucuronide, which adds a high sweetness that is roughly 941 times sweeter than sugar.

- Food flavor for tobacco:

Apparently, licorice is used by tobacco manufacturers at various phases of production to flavour the tobacco and sweeten the smoke. The smoke has a mellow, sweet, and woodsy flavour thanks to the licorice.Up to 4% of the total weight of the tobacco used in one cigarette can be made up by the licorice that is added to it. Other tobacco products like cigars and chewing tobacco also have licorice as a flavouring.

The tobacco business makes extensive use of licorice extracts. Licorice gives tobacco a moderate flavour and scent in addition to its sweet flavour. Additionally, it stops the tobacco from drying out. The American corporation MAFCO provides the licorice extracts used in the tobacco industry. In 1987, Japan imported licorice from China and the USA; in 2007, it came from the USA, Israel, China, and India. The trade data for licorice extracts in Japan between 1987 and 2007 is available. Licorice extracts for the manufacturing of tobacco were heavily imported into Japan from the USA (848,704 kg in 1987 and 458,179 kg in 2007), with a value of more than 500 million.

- Confectionary:

Licorice is a black-colored confection that is usually flavoured and coloured with Glycyrrhiza glabra root extract. Around the world, a wide range of licorice sweets are produced. In North America, black licorice is distinguished from similar confectionery varieties that are not flavoured or coloured black with licorice extract but are commonly manufactured in the form of similarly shaped chewy ropes or tubes and are commonly referred to as red licorice. Black licorice, along with anise extract, is a popular flavour in other types of confectionery, such as jellybeans. In addition to these, licorice allsorts and other licorice-based sweets are available in the United Kingdom. In addition to the typical sweet variations found in the United Kingdom as Licorice allsorts.

In addition to the sweet varieties found in the United Kingdom and North America, Dutch and Nordic licorice contains ammonium chloride rather than sodium chloride, most notably in salty licorice, which has a strong salty rather than sweet flavour. Licorice extract, sugar, and a binder are the three main ingredients in black licorice confectionery. The base is usually starch/flour, gum arabic, gelatin, or a combination of these ingredients. Extra flavouring, beeswax for a shiny surface, ammonium chloride, and molasses are added ingredients. Ammonium chloride, in concentrations of up to 8%, is primarily used in salty licorice candy. Even regular licorice candy, on the other hand, can contain up to 2% ammonium chloride, the taste of which is less noticeable due to the higher sugar concentration.

Other uses:

- Licorice drinks:

An extract obtained for consuming the licoricedrink is prepared in Turkey, Southeastern Anatolia, particularly in Adana, Hatay, Sanliurfa, Diyarbakir, Batman, Mardin, Adyaman, Siirt, and Gaziantep, by mixing roots with water. It is widely used in these areas. The fresh roots are sun-dried after being carefully washed with water to remove any adherents. To obtain a fiber-like product, dried roots are crushed. These fibres are mixed with carbonate and cinnamon powder for a pleasant aroma. It is placed in a large wooden bowl with a bottom tap and water is added. Water is tapped from the wooden bowl and poured over the fibre repeatedly until the extract has the desired taste and colour.

There are two types of licoriceavailable in the market: roots and extract. The long brown roots are typically harvested from the field and sun-dried from plants that are 3 to 4 years old. The fibrous roots and residual stems are separated. The extract is made by chopping the roots and then steam extracting them, and the liquor is filtered and concentrated. The end result could be a solid block or a spray-dried powder.

Boil 1 cup of water and pour it over one licorice root to make licorice root tea. Allow 5 minutes for the licorice root to steep. Serve with a strainer.Licorice root tea is commonly used to treat gastrointestinal (stomach) issues and to soothe a sore throat.

Licorice root is a popular tea ingredient that gives loose leaf tea blends a sweet flavour. Apart from being delicious, licorice root has a plethora of health benefits, such as reducing inflammation, aiding digestion, and boosting the immune system. We'll go over the benefits of licorice root tea, as well as some licorice root blends to consider.

Figure 16: licorice herbal tea

- Use as animal feed:

Over the last decade, the use of antibiotics as feed additives in livestock production has been restricted globally. The restriction is in response to concerns about the spread of antibiotic-resistant pathogenic bacteria and antibiotic residues in animal product. This has heightened interest in alternative phytogenic feed additives for improving livestock health and productivity. One possible option is to use licorice root extract as a feed additive, which has been studied primarily in monogastrics over the last 15 years and only sparingly in ruminants.

Ozhan and Gol (1975) demonstrated that the leaves of G. glabra can be used to meet the forage needs of ruminant animals when forage quality and quantity are limited. Despite the fact that G. glabra is one of the naturally grown plants used to meet the needs of small ruminants in some semiarid regions of Turkey during critical times of the year, there have been very few studies to validate the nutritional quality of G. glabra leaves.

Zhang et al. (2015) recently demonstrated that adding 0.4% licorice root extract (16.4% total flavonoids content) to a sheep diet resulted in higher antioxidant capacity of meat compared to the non-supplemented group. Furthermore, Sajjadi et al. (2014) demonstrated that supplementing Holstein heifer calves with 10 g day1 of licorice root extract (equivalent to 0.56% of the diet) increased total immunoglobulin (Ig) concentration and thus improved their immunological status. Kim et al. (2013), on the other hand, demonstrated that supplementing the diet with 0.5% licorice (no licorice chemical composition was reported) had no effect on dry matter intake (DMI) and average daily gain (ADG) of Hanwoo steers.

Figure 17: animal feed usage

- Plant as a dye:

. Different colours were obtained from licoriceroot when no mordant was used, when the mordant was used alone, and when the potassium bichromate mordant was kept constant and the two mordants were mixed. Open straw yellow, open henna green, honey colour, milky coffee, chick yellow, green yellow, dirty yellow, dry oak leaf, open dirty yellow, open earth, hay yellow, open quince feather, dark quince feather, open dry oak leaf, and dark dry oak leaf are some of the colours available. These are the most common colours used in hand-woven carpets and rugs. These are highly recommended for use in eco-friendly herbal dyeing of rugs and carpets.

The present invention is directed to a natural dyeing method of fabric using licorice extract or glycyrrhizin separated from it. More specifically, the natural dyeing method of the fabric

a) Removing impurities contained in the fabric and refining;

b) pulverizing licorice to prepare a licorice salt solution containing licorice extract, which is extracted with an extraction solvent selected from the group consisting of distilled water, ethanol, and mixtures thereof;

c) Embedding using a mordant selected from the group consisting of aluminum mordant, copper mordant, and

d) Using a licorice salt solution to dye the fabric

Natural dyeing using licorice extract with various functionalities or glycyrrhizin separated from it, according to the present invention, can not only apply useful properties of licorice to the fabric, but also increase the amount of dyeing, antibacterial, antiperspirants, and UV protection. There is a benefit that can improve capabilities, such as sex.

After 20 minutes of US treatment, licorice bark extract demonstrated excellent colour depth (K/S) onto cotton fabric at 65°C. Using bio-mordants, it was discovered that acacia extract, turmeric extract, and henna extracts had exceptional colour strength.

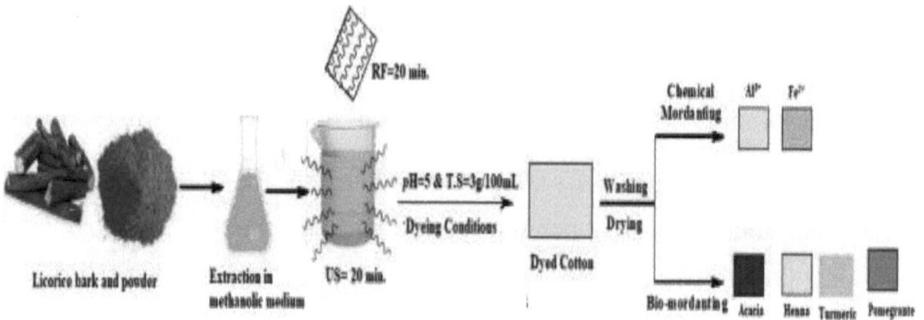

Figure 18: dye production

Antiviral activity:

Several virus diseases, including chronic hepatitis B and C, as well as human acquired immunodeficiency syndrome (AIDS), have been treated as potential therapeutic agents with licorice and glycyrrhizate compounds derived from licorice extract. Many studies have found that glycyrrhizin and glycyrrhizic acid have antiviral activity, inhibiting the growth and cytopathology of hepatitis A and C, as well as the immunodeficiency virus (HIV). Glycyrrhizin and its derivatives from G. glabra have been shown to reduce hepatocellular damage in chronic hepatitis B and C, as well as to have antiviral activity against HIV-1, SARS-related coronavirus, respiratory syncytial virus, arboviruses, vaccinia virus, and vesicular stomatitis virus. Glycyrrhizin also has an antiviral effect, either by inhibiting viral particle-to-cell membrane binding or by inhibiting cellular signal transduction mechanisms

18-glycyrrhetinic acid has been identified as a promising biological substitute for the potential therapy of vulvovaginal candidiasis. In vitro antiviral effects for viruses resulting respiratory infections, such as influenza virus and the SARS coronavirus, as well as human immunodeficiency virus (HIV), have been reported.

To date, 73 bioactive components and 91 potential targets have been identified among the components isolated from licorice14, 15. Many studies have shown that the antiviral activity is mediated by two triterpenoids, GL16, 17, and GA18. The possible virus prevention mechanisms for GL and GA, as well as the viral types, are shown in following figure.

Component	Antiviral mechanism	Viral type
GL	Affect release step while infectious HCV particles are infecting cells.	HCV
	Inhibit HCV full length viral particles and HCV core gene expression.	
	Reduce adhesion force and stress between CCEC and PMN.	HSV
	Block the degradation of nuclear factor κB inhibitor IκB.	CVB3
	Activate T lymphocyte proliferation.	DHV
	Weaken H5N1-induced production of CXCL10, IL-6 and CCL5, and suppress H5N1-induced apoptosis.	H5N1
	Reduce HMGB1 binding to DNA, and inhibit influenza virus polymerase activity.	Influenza virus
	Inactivate CVA16 directly, while the effect of anti-EV71 is associated with an event(s) during the virus cell entry.	CVA16 EV71
	Establish a resistance state to HSV1 replication.	HSV1
GA	Reduce the levels of viral proteins VP2, VP6 and NSP2 at a step or steps subsequent to virus entry.	Rotavirus

Figure 19: antiviral activity compounds

G. uralensis roots have yielded 13 new oleanane-type triterpenoid saponins, uralsaponins M-Y (1-13), and 15 known analogues (14-28). Extensive NMR and MS data analysis was used to identify the structures of 1-13. Following hydrolysis, the sugar residues were identified using gas chromatography and ion chromatography in conjunction with pulsed amperometric detection. Saponins containing galacturonic acid (1-3) or xylose (5) residues have been reported for the first time from Glycyrrhiza species. Compounds 1, 7, 8, and 24 have been shown to have good inhibitory activities against the influenza virus A/WSN/33 (H1N1) in MDCK cells, with IC50 values of 48.0, 42.7, 39.6, and 49.1 M, respectively, compared to the positive control oseltamivir phosphate, which has an IC50 value of 45.6 M.

Furthermore, compounds 24 and 28 demonstrated anti-HIV activity with IC50 values of 29.5 and 41.7 M, respectively (Song et al. 2014). According to Adianti et al. (2014), glycycoumarin, glycyrin, glycyrol, and liquiritigenin isolated from G. uralensis, as well as isoliquiritigenin, licochalcone A., and glabridin, can be good candidates for seed compounds in the development of antivirals against HCV. Following in vivo studies, glycyrrhizin exhibits a strong lymphocytic proliferation response in white Pekin ducklings and a good immune stimulant and antiviral effect against DHV (Soufy et al. 2012). A combination of glutamyl-tryptophan and glycyrrhizinin has a protective effect on the death of H3N2 virus-infected mice.

- Anti-inflammatory effect:

Matsui et AL report's clearly shows that licorice species have been used to treat many allergies and other inflammatory diseases. Glycyrol (benzofuran coumarin) isolated from G. uralensis has been extensively studied for its anti-inflammatory properties. They discovered that glycerols may have anti-inflammatory properties. The anti-inflammatory activity of licoriceroot constituents is similar to that of hydrocortisone, according to Vibha et al. Their explanation for this discovery is that it occurs due to inhibition of phospholipase A2 activity, an enzyme important in many inflammatory processes. Glycyrrhetinic acid (ED 50, 200 mg kg1) has been shown to inhibit carrageenan-induced rat paw edoema and to have anti-allergic activity. Secondary metabolites of G. glabra include glycyrrhizic acid, glabridin, and licochalcone A also show an anti-inflammatory effect.

Licorice appears to be an excellent alternative choice for the treatment of inflammation in epidemic diseases affecting populations, particularly children. Anti-inflammatory compounds include glabridin, isoliquiritigenin, dehydroglyasperin C, licoricidin, licorisoflavan A, echinatin, and glyurallin B. Postischemic brain with middle cerebral artery occlusion, LPS-induced BV-2 microglia (Kim and Fu), childhood atopic dermatitis, allergic airway inflammation, and TPA-induced mouse ear edoema were all studied.

In the absence of an appropriate pharmacological intervention in Western medicine to treat COVID-19 patients, the goal of this review is to highlight the potential of a medicinal plant species known as licorice, which belongs to the shrub category and whose phytochemicals have antiviral and anti-inflammatory properties, and to recommend that clinical trials be conducted to assess its potency in treating COVID-19 symptoms. Many other medicinal plants, including Sambucus nigarac, Desmodium canadense, the Lamiaceae family, Asteraceae, Geraniaceae, and others, have phytochemicals with similar properties. However, licorice has been explicitly suggested due to its: I significant antiviral property against several viruses, including SARS-CoV, (ii) strong anti-inflammatory property, which has been observed in many rat model studies, (iii) autophagy-enhancing mechanism, (iv) established use in Chinese and Japanese medicine, and (v) established use in Chinese and Indian Ayurvedic medicines, and (v) wide distribution.

- Antimicrobial and antifungal activity:

Antibiotic resistance has created an urgent need for alternative treatments to treat diseases. Many studies in recent years have shown that licorice aqueous extract, ethanol extract, and supercritical fluid extract have potent inhibitory effects on Gram-positive and Gram-negative bacteria, including Staphylococcus aureus, Escherichia coli, Pseudomonas aeruginosa, Candida albicans, and Bacillus subtilis. These extracts are also being considered as potential synthetic fungicide alternatives or as lead compounds for new classes of synthetic fungicides. Based on its antibacterial properties, licorice may be used as an alternative therapy for treating dental caries, periodontal disease, digestive anabrosis, and tuberculosis. The potential antimicrobial mechanisms of the active ingredients and microorganism types are shown in following figure

Component	Antimicrobial mechanism	Microbial type
GA	Decrease the expression of *SaeR* and *Hla*, which are the key virulence genes of MRSA.	*S. aureus*
Exert the Th1-immunological adjuvant activity.	*C. albicans*	
LCA	Inhibit the biofilm formation and prevent yeast-hyphal transition.	*C. albicans*
LCE	Reduce the production of α-toxin.	*S. aureus*
GLD	Prevent yeast-hyphal transition.	*C. albicans*
LTG	Decrease the production of α-hemolysin.	*S. aureus*

Figure 20: antimicrobial and antifungal activity

Recently, a number of clinical and pharmaceutical studies have been conducted that show how licorice aqueous, ethanol, and supercritical fluid extract have antibacterial properties. Since many years ago, this activity in plant oils and extracts has been acknowledged. It has been suggested that alkaloids, saponins, flavonoides, tannin, glycosides, and phenols are responsible. Meghashri has also demonstrated the inhibition by ethanolic extracts of G. glabra and their fractions. The methanolic extracts of G. glabra, however, appear to have a stronger fungicidal impact against Arthrinium sacchari and Chaetomium funicola. G. glabra extracts have been shown to have antifungal action by Tharkar as well. Similar information about the antibacterial activity of G. glabra against Mycobacterium tuberculosis has been published by Gupta et al.

Alcohol, chloroform, and acetone-based licorice extracts exhibit antibacterial activity against Escherichia coli, Stapphylococcus aureus, and Bacillus subtilis. G. glabra extract has been shown to have a strong antibacterial effect against Propionibacterium acnes and Staphylococcus epidermidis. Varsha et al. have shed insight on the antibacterial activity of

G. glabra extract against Pseudomonas aeruginosa, Escherichia coli, Staphylococcus epidermidis, S. aureus, and Baillus subtilis. G. glabra extracts have been shown to have antibacterial activity against Pseudomonas aeruginosa and B. subtilis. According to Tanaka et al (2001). Study of upper respiratory tract bacteria such Streptococcus pyogenes, Haemophilus influenza, and Moraxella catarrhalis, secondary metabolites derived from Glycyrrhiza species exhibit antibacterial action against these.

Glabridin, an active component of Glycyrrhiza glabra roots, has been shown to be effective against filamentous fungus as well as yeast. At a minimum inhibitory dosage of 31.25-250 microg/mL, glabridin also demonstrated resistance-modifying effect against drug-resistant mutants of Candida albicans. This is the first report of the compound's efficacy against drug-resistant mutants, despite the fact that it has previously been shown to be effective against Candida albicans.

According to Yang et al. (2015), licorice shows excellent potential for use as an alternative medication in the treatment of TB, digestive anabrosis, dental caries, and periodontal disease. Additionally, it has the potential to serve as a lead molecule for new classes of synthetic fungicides or as an alternative to synthetic fungicides.

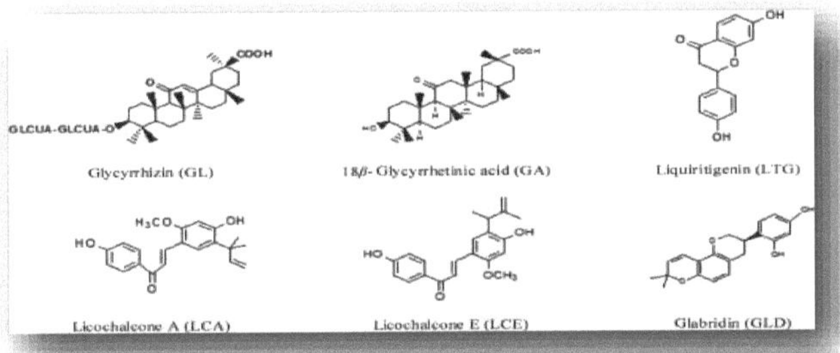

Figure 21: structures of antimicrobial components

- Antidepressants:

The mice immobility tests, according to Dhingra and Sharma, have unequivocally shown that licoriceis quite effective as an antidepressant as well as in the memory-enhancing action. Following research on depressed rats, Zhao et al. (2006) conducted studies that showed liquiritin and several other secondary metabolites from G. uralensis have an antidepressant impact on chronic stress. Data on the antidepressant-like action of liquiritin and isoliquiritin in the forced swimming test (FST) and tail suspension test (TST) in mice were also published by Wang et al. (2008). Recent studies have suggested that the mechanism of action of these substances may involve an increase in 5-hydroxytryptamine and norepinephrine in the mouse cortex, hypothalamus, and hippocampus.

Gareri et al. have reported that a different component in licorice, carbenoxolone, had sedative and muscle relaxing properties in mice. These mice are genetically predisposed to epilepsy (GEPRs) (2004). According to research on the effects of G. glabra root extract on learning and memory in a different experiment, 150 and 225 mg/kg dosages effectively improve learning and memory compared to the control. The antioxidant and anti-inflammatory properties of plant extract may be the cause. In these conditions, less oxidative stress is applied to the vulnerable brain cells, reducing brain damage and enhancing neural function.

Numerous researchers have looked into how licorice aqueous extract affects memory and learning. One of these studies was conducted by Chakravarthi and Avadhani, who looked at how the extract affected the dendritic structure of hippocampal cornu ammonis area three (CA3) neurons. These researchers claim that dendritic arborization and dendritic crossings in hippocampus pyramidal neurons are clearly enhanced at dosages of 150 and 225 mg/kg. This has proven to have properties that stimulate neuronal dendritic development. Additionally, these researchers discovered that all doses of this extract considerably improve memory. The use of 150 and 225 mg/kg dosages greatly improves memory and learning. These extracts have nootropic and antiamnestic properties, which are brought about by increasing monoaminergic transmission in the cortex, hippocampus, and striatum.

Glabridin and 2, 2, 4-trihydroxychalcone are among a huge number of substances extracted from licorice that are said to enhance memory and learning. These are frequently employed in the management of conditions affecting the cardiovascular and neurological systems. The higher doses of the former substance successfully combat scopolamine-induced amnesia. According to Hasanein (2011), rat trials show that this substance also protects against the negative effects of diabetes on memory and learning. For the purpose of finding new anti-disease Alzheimer's medications, inhibition of the beta-site amyloid precursor protein cleaving enzyme 1 is thought to be a successful approach. Inhibitor 2, 2,4-trihydroxychalcone cleaving enzyme 1 of the novel beta-site amyloid precursor protein has been evaluated to lessen mouse impairment. All of these data highlight the fact that licorice appears to be a promising medication for enhancing memory, preventing learning disabilities, treating Alzheimer's disease and other neurodegenerative disorders, and dementia.

This shows that the licorice extract's antidepressant-like effects appear to be mediated by an increase in brain norepinephrine and dopamine, but not serotonin. The antidepressant-like effects of licorice may benefit from its monoamine oxidase inhibitory effect. Therefore, it can be said that licorice extract may have an effect similar to an antidepressant.

- Inhibitory effect on diabetes:

Diabetes mellitus is a collection of metabolic illnesses brought on by flaws in insulin secretion and/or activity that are characterised by hyperglycemia and eventually glycosuria. Patients with diabetes mellitus experience not only physical and emotional pain, but also certain significant, potentially fatal complications.

One of the Chinese herbs that is most frequently utilised is licorice. It is frequently used with other herbs to treat metabolic problems, particularly diabetes mellitus, in TCM treatment. For instance, licorice is present in a number of decoctions that are frequently used to treat people with diabetes mellitus, including Xiaoyaosan, Gegenqinlian, Yitangkang, and KIOM-79. In Wistar rats, Xiaoyaosan Decoction reduces mood disorders, alters neuronal plasticity impairment, and lowers blood sugar levels5. In RIN-5F cells, Gegenqinlian Decoction controls glucose homeostasis, promotes endogenous insulin production, and reduces insulin resistance. The metabolic issues connected to type II diabetes can be resolved with yitangkang.

To prevent the onset of diabetes, licorice blocks a number of pathological and physiological changes brought on by d-galactose, including insulin resistance and oxidative stress/free radical damage. Glycyrrhizin, glycyrrhetinic acid, liquiritigenin, isoliquiritigenin, glabridin, licochalcone A, licochalcone E, and several other flavonoids are just a few of the substances extracted from licorice that have been found to have an inhibitory effect on diabetes. Through suppressing Akt activation and transforming growth factor- signalling, licoriceextract can be a highly effective therapeutic agent for the prevention and treatment of diabetes nephropathy, which is characterised by mesangial fibrosis and glomerulosclerosis. It lowers blood sugar levels, improves renal function, slows weight loss, and lessens the negative effects of diabetes on renal glutathione, malondialdehyde, and catalase and superoxide dismutase activity. Due to its antioxidant and antihyperglycemic properties, it also improves the overall antioxidant capacity of diabetic kidneys and may have therapeutic benefits for diabetes. According to all of these reports, licorice is a highly effective medicinal drug for the treatment of diabetes.

- Antioxidant effect:

Reactive oxygen species (ROS) generation and antioxidant defence capacity are out of balance, which leads to oxidative stress. Superoxide anions, singlet oxygen, nitric oxide, peroxynitrite, hydrogen peroxide, and hydroxyl radicals are just a few examples of the ROS that can cause chronic illnesses including cancer, ageing, as well as neuro-degenerative and cardiovascular diseases by harming lipids, DNA, and proteins.

The information presented in a 2011 study by Siracusa et al. makes it abundantly evident that G. glabra leaf extract possesses antioxidant, anti-genotoxic, and anti-inflammatory properties. The roots of Glycyrrhiza have been shown to have a variety of phytochemical components. The isoflavones glabridin and hispaglabridins A and B exhibit considerable antioxidant activity, making them a possible source of antioxidants. The flavonoids luteolin, rutin, and apigenin contained in the roots of G. glabra exhibit notable antioxidant properties. Phenolic molecules are mentioned as the major substance connected to antioxidant activity. In hypoxic rats, the ethanol extract of G. glabra has been shown to have a cerebroprotective effect; this action may be mediated by the plant's antioxidant properties. Additionally, 400 g/ml of G. glabra essential oil has 85.2% DPPH radical scavenging activity. In contrast, a dose of 62.5 g of methanolic extract exhibits 91.3% scavenging activity. Licochalcone C is said to have antioxidant properties because it lowers superoxide radical generation, which lowers inducible nitric oxide synthase activity.

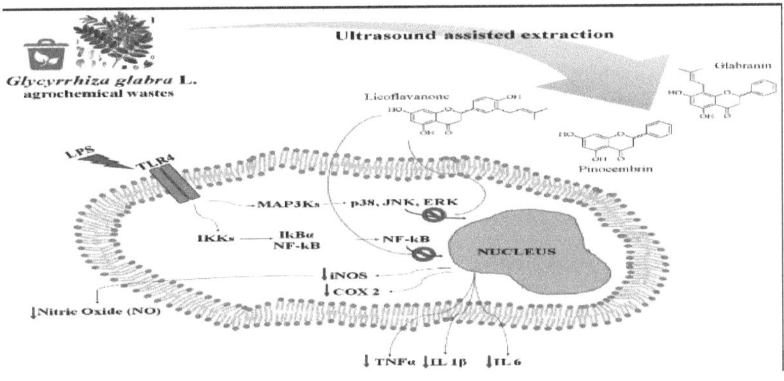

Figure 22: antioxidant effect

Side effect of licorice:

Although licorice's harmful effects are well known, its potentially poisonous ingredients have not yet been fully verified. Due to the hypertension brought on by its excessive usage, the use of this plant is contraindicated in people with high blood pressure. When combined with stimulant laxatives or hypotensive diuretics, this plant should not be utilized (such as thiazides).

Some research on licorice claim that the TA98 and TA100 strains did not respond favorably to ethanol extracts of G. glabra that included glycyrrhizin and glycyrrhetic acid. Using test systems for chromosomal aberration, micronucleus, and the Ames IITM, the genotoxic potential of G. glabra has been examined. The findings show that this plant doesn't have any mutagenic properties, metabolic activation or not. According to reports, TA98 is more susceptible to the mutagens found in licorice water extract. At 0.24 mg/mL, licorice extract's cytotoxic activity is 63%; higher percentages become visible as concentrations rise to 4.8 mg/ml. Chloroform, methanol, and water extracts of G. glabra have been shown to have IC50 values of 0.45, 0.99, and 1.29 M in the breast MCF7 cell line, respectively. G. glabra water extract has mutagenic activity against the TA100 strain in the presence of metabolic activation, but only at the maximum extract concentration, according to Abudayyak et al. According to reports, this plant extract can occasionally cause an acquired type of apparent mineralocorticoid excess syndrome, which manifests as salt retention, hypokalemia, and elevated blood pressure.

In TCM, licorice is typically listed as a harmless herb. Edema and the apparent mineralocorticoid excess syndrome are the results. Due to licorice-induced hyperaldosteronism, excessive daily consumption may result in low potassium levels, which can then lead to hypertension and heart disease. By ceasing to consume it and its products, the hypertension that their use caused can be treated.

Licorice may make fetal abnormalities and weight loss worse by up regulating cytochrome P450 type 2B. However, according to Korean experts, there is no increased risk of stillbirth among women who take licorice for cold and cough symptoms. Additionally, its extract raises human estrogen levels and may raise the risk of estrogen-mediated cancer.

To summarize following are side effects of licorice:

- o Lack of a menstrual cycle
- o Reduced sexual desire, erection complications
- o Fluid buildup in the lungs (pulmonary edema)
- o Sodium and fluid retention
- o Headache, Tiredness, Weakness
- o Elevated blood pressure (hypertension)
- o Cerebral hypertension, occasionally, healthy individuals have brain damage
- o Low amounts of potassium (hypokalemia)
- o Paralysis, Swelling

References:

1. "Glycyrrhiza glabra". Germplasm Resources Information Network (GRIN). Agricultural Research Service (ARS), United States Department of Agriculture (USDA). Retrieved 6 March 2008.

2. ^ "The Plant List: A Working List of All Plant Species". Retrieved 2017-03-07.

3. ^ "Glycyrrhiza pallida Boiss., Diagn. Pl. Orient. ser. 2, 2: 22 (1856)". The International Plant Names Index. Retrieved 2017-03-07.

4. https://tomsgroup.com/en/toms-history/the-story-of-liquorice/#:~:text=The%20history%20of%20liquorice%20can,plant%20that%20rejuvenated%20ageing%20men.

5. Md. KamrulHasana1IffatAraa1Muhammad Shafiul AlamMondalbYearulKabirb https://www.sciencedirect.com/science/article/pii/S2405844021013438#:~:text=This%20medicinal%20plant%20is%20found,several%20Glycyrrhiza%20species%20%5B8%5D.

6. Afchar D, Cave A, Vaquette J. Etudes des reglisses d'Iran. I. Flavonoides de Glycyrrhiza glabra var. glandulifera. Plant Med Phytother. 1980;14(1):46–50. [Google Scholar]

7. https://www.ncbi.nlm.nih.gov/pmc/articles/PMC7122586/

8. https://www.sciencedirect.com/science/article/abs/pii/S0378874122006523

9. http://bioweb.uwlax.edu/bio203/s2012/olsen_bran/classification.htm

10. https://www.ncbi.nlm.nih.gov/pmc/articles/PMC7167772/

11. Wang, Q. , Qian, Y. , Wang, Q. , Yang, Y.-f. , Ji, S. , Song, W. , … Ye, M. (2015). Metabolites identification of bioactive licorice compounds in rats. Journal of Pharmaceutical and Biomedical Analysis, 115, 515–522. [PubMed] [Google Scholar]

12. Rizzato, G. , Scalabrin, E. , Radaelli, M. , Capodaglio, G. , & Piccolo, O. (2017). A new exploration of licorice metabolome. Food Chemistry, 221, 959–968. [PubMed] [Google Scholar]

13. Yu, J. Y. , Ha, J. Y. , Kim, K. M. , Jung, Y. S. , Jung, J. C. , & Oh, S. (2015). Anti-inflammatory activities of licorice extract and its active compounds, glycyrrhizic acid, liquiritin and liquiritigenin, in BV2 cells and mice liver. Molecules, 20(7), 13041–13054. [PMC free article] [PubMed] [Google Scholar]

14. Albermann, M. E. , Musshoff, F. , Hagemeier, L. , & Madea, B. (2010). Determination of glycyrrhetic acid after consumption of liquorice and application to a fatality. Forensic Science International, 197(1), 35–39. [PubMed] [Google Scholar]

15. Fukui, H. , Goto, K. , & Tabata, M. (1988). Two antimicrobial flavanones from the leaves of Glycyrrhiza glabra . Chemical & Pharmaceutical Bulletin, 36(10), 4174–4176. [PubMed] [Google Scholar]

16. Simmler, C. , Pauli, G. F. , & Chen, S.-N. (2013). Phytochemistry and biological properties of glabridin. Fitoterapia, 90, 160–184. [PMC free article] [PubMed] [Google Scholar]

17. Chouitah, O. , Meddah, B. , Aoues, A. , & Sonnet, P. (2011). Chemical composition and antimicrobial activities of the essential oil from Glycyrrhiza glabra leaves. Journal of Essential Oil-Bearing Plants, 14(3), 284–288. [Google Scholar]

18. https://licorice.com/blogs/news/learn-about-the-different-types-of-licorice

19. https://www.wise-geek.com/what-is-sour-licorice.htm

20. Anilkumar DI, Hemang J and Nishteswar K (2012). Review of Glycyrrhiza glabra (yastimadhu) - abroad spectrum herbal drug. An. Int. J. Pharm. Sci., 3195

21. Bone K (1990). Liquorice the universal herb. The British J. Phytotherapy, 1(2):7-13.

22. Mohammad N and Rehman S (1985). Performance of Glycyrrhiza glabra in Mastung valley, Baluchistan. Pak. J Agric. Res., 6(3): 176-179.

23. Mori K, Sakai H, Suzuki S, Akutsu Y, Ishikawa M, Imaizumi M, Tada K, Aihara M, Sawada Y and Yokoyama M et al. (1990). Effects of glycyrrhizin (SNMC: Stronger neo minophagenin hemophilia patients with HIV-1 infection. Tohoku J. Exp. Med., 162(2): 183-193.

24. https://plantura.garden/uk/herbs/liquorice/liquorice-overview

25. https://www.reviewgeek.com/59649/these-apps-will-help-you-keep-all-of-your-new-houseplants-alive/

26. https://www.bhg.com/gardening/plant-dictionary/annual/licorice-plant/#:~:text=When%20looking%20for%20a%20home,though%20it%20prefers%20regular%20watering.

27. https://www.homestolove.com.au/licorice-plant-11335

28. https://www.urbangardengal.com/harvest-dry-licorice-roots/#:~:text=Air%20drying%20is%20a%20popular,six%20weeks%20until%20fully%20dry

29. https://www.google.com/url?sa=i&url=https%3A%2F%2Fperfumesociety.org%2Fing
redients-
post%2Flicorice%2F&psig=AOvVaw16wmjlIyO2_ZMGPGuHJNTa&ust=16648751
35954000&source=images&cd=vfe&ved=0CA0QjhxqFwoTCIDH5bPdw_oCFQAA
AAAdAAAAABAE

30. licorice use for canker sores - Search (bing.com)

31. Lysine for Canker Sores | New Health Advisor

32. Licorice for Canker Sores - Using tea Tree Oil and Other herbs for mouth sores
(teatreewonders.com)

33. https://www.verywellhealth.com/the-benefits-of-licorice-root-89727

34. https://pubmed.ncbi.nlm.nih.gov/17978499/#:~:text=The%20administration%20of%2
0the%20licorice,activity%20of%20the%20licorice%20extract.

35. https://ars.els-cdn.com/content/image/1-s2.0-S0378874120335236-fx1_lrg.jpg

36. https://www.sciencedirect.com/science/article/abs/pii/S0378874120335236

37. Functional Dyspepsia

38. Licorice Extract for Skin: The Complete Guide (byrdie.com)

39. https://deputyprimeminister.gov.mt/en/environmental/Pages/Policy-Coordinating-
Unit/PITOC/Licorice.aspx

40. https://www.ncbi.nlm.nih.gov/pmc/articles/PMC7124151/

41. https://www.mdpi.com/2079-9284/9/1/7/htm

42. https://pubmed.ncbi.nlm.nih.gov/11312777/

43. https://deputyprimeminister.gov.mt/en/environmental/Pages/Policy-Coordinating-
Unit/PITOC/Licorice.aspx#:~:text=Tobacco%20manufacturers%20reportedly%20use
%20licorice,tobacco%20used%20in%20one%20cigarette.

44. https://en.wikipedia.org/wiki/Liquorice_(confectionery)#:~:text=Liquorice%20(Britis
h%20English)%20or%20licorice,are%20produced%20around%20the%20world.

45. https://www.webmd.com/diet/licorice-root-tea-is-it-good-for-you#1

46. https://www.artfultea.com/tea-wisdom-1/licorice-root-tea-benefits

47. https://www.mdpi.com/2076-2615/9/5/279/htm

48. Arlı M, Kayabaşı N, Kızıl S (2002) A research on the colours obtained from licorice
(Glycyrrhiza glabra L.) plant in natural dyeing and their some fastnesses. Tarım
Bilimleri Dergisi 8(3):227–231

49. https://link.springer.com/article/10.1007/s11356-021-18472-5

50. https://www.sciencedirect.com/science/article/pii/S2211383515000799

51. Smirnov VS, Zarubaev VV, Anfimov PM, Shtro AA. Effect of a combination of glutamyl-tryptophan and glycyrrhizic acid on the course of acute infection caused by influenza (H3H2) virus inmice. Vopr Virusol. 2012;57:23–27. [PubMed] [Google Scholar]

52. https://www.mdpi.com/2223-7747/10/12/2600/htm

53. Luo L, Jin Y, Kim ID, Lee JK. Glycyrrhizin attenuates kainic Acid-induced neuronal cell death in the mouse hippocampus. Exp. Neurobiol. 2013;22:107–115. doi: 10.5607/en.2013.22.2.107. [PMC free article] [PubMed] [CrossRef] [Google Scholar] [Ref list]

54. Kim HJ, Lim SS, Park IS, Lim JS, Seo JY, Kim JS. Neuroprotective effects of dehydroglyasperin C through activation of heme oxygenase-1 in mouse hippocampal cells. J Agric Food Chem. 2012;60:5583–5589. doi: 10.1021/jf300548b. [PubMed] [CrossRef] [Google Scholar] [Ref list]

55. https://www.ncbi.nlm.nih.gov/pmc/articles/PMC4629407/

56. https://pubmed.ncbi.nlm.nih.gov/19170157/

57. Statti GA, Tundis R, Sacchetti G, Muzzoli M, Bianchi A, Menichini F. Variability in the content of active constituents and biological activity of Glycyrrhiza glabra. Fitoterapia. 2004;75:371–374. doi: 10.1016/j.fitote.2003.12.022. [PubMed] [CrossRef] [Google Scholar] [Ref list]

58. Shinwari ZK, Khan I, Naz S, Hussain A. Assessment of antibacterial activity of three plants used in Pakistan to cure respiratory diseases. Afr J Biotechn. 2009;8(24):7082–7086. [Google Scholar] [Ref list]

59. Yang R, Wang LQ, Yuan BC, Liu Y. The pharmacological activities of licorice. Planta Med. 2015;81(18):1654–1669. doi: 10.1055/s-0035-1557893. [PubMed] [CrossRef] [Google Scholar] [Ref list]

60. https://www.sciencedirect.com/science/article/pii/S2211383515000799

61. Dhingra D, Sharma A. Antidepressant-like activity of Glycyrrhiza glabra L. in mouse models of immobility tests. Prog Neuro-Psychopharmacol Biol Psychiatry. 2006;30:449–454. doi: 10.1016/j.pnpbp.2005.11.019. [PubMed] [CrossRef] [Google Scholar] [Ref list]

62. Gareri P, Condorelli D, Belluardo N, Russo E, Loiacono A, Barresi V, Trovato-Salinaro A, Mirone MB, Ferreri Ibbadu G, De Sarro G. Anticonvulsant effects of carbenoxolone in genetically epilepsy prone rats (GEPRs) Neuropharmacology.

2004;47(8):1205–1216. doi: 10.1016/j.neuropharm.2004.08.021. [PubMed] [CrossRef] [Google Scholar] [Ref list]

63. Michel HE, Tadros MG, Abdel-Naim AB, Khalifa AE. Prepulse inhibition (PPI) disrupting effects of Glycyrrhiza glabra extract in mice: a possible role of monoamines. Neurosci Lett. 2013;544:110–114. doi: 10.1016/j.neulet.2013.03.055. [PubMed] [CrossRef] [Google Scholar] [Ref list]

64. Cui YM, Ao MZ, Li W, Yu LJ. Effect of glabridin from Glycyrrhiza glabra on learning and memory in mice. Planta Med. 2008;74:377–380. doi: 10.1055/s-2008-1034319. [PubMed] [CrossRef] [Google Scholar] [Ref list]

65. https://pubmed.ncbi.nlm.nih.gov/16443316/#:~:text=This%20suggests%20that%20antidepressant%2Dlike,to%20the%20antidepressant%2Dlike%20activity.

66. https://www.sciencedirect.com/science/article/abs/pii/S0378874120330981

67. Wu F, Jin Z, Jin J. Hypoglycemic effects of glabridin, a polyphenolic flavonoid from licorice, in an animal model of diabetes mellitus. Mol Med Rep. 2013;7:1278–1282. doi: 10.3892/mmr.2013.1330. [PubMed] [CrossRef] [Google Scholar] [Ref list]

68. Yang R, Wang LQ, Yuan BC, Liu Y. The pharmacological activities of licorice. Planta Med. 2015;81(18):1654–1669. doi: 10.1055/s-0035-1557893. [PubMed] [CrossRef] [Google Scholar] [Ref list]

69. Hazra B, Biswas S, Mandal N. Antioxidant and free radical scavenging activity of Spondias pinnata. BMC Complement Altern Med. 2008;8:63. [PMC free article] [PubMed] [Google Scholar] [Ref list]

70. Lateef M, Iqba L, Fatima N, Siddiqui K, Afza N, Zia-ul-Haq M, Ahmad M. Evaluation of antioxidant and urease inhibition activities of roots of Glycyrrhiza glabra. Pak. J. Pharm. Sci. 2012;25:99–102. [PubMed] [Google Scholar] [Ref list]

71. Franceschelli S, Pesce M, Vinciguerra I, Ferrone A, Riccioni G, Antonia P, et al. Licocalchone-C extracted from Glycyrrhiza glabra inhibits lipopolysaccharide-interferon-γ inflammation by improving antioxidant conditions and regulating inducible nitric oxide synthase expression. Molecules. 2011;16(7):5720–5734. doi: 10.3390/molecules16075720. [PMC free article] [PubMed] [CrossRef] [Google Scholar] [Ref list]

72. https://www.mdpi.com/2076-3921/8/6/186

73. https://www.rxlist.com/consumer_licorice/drugs-condition.htm

74. Choi JS, Han JY, Ahn HK, Ryu HM, Kim MY, Chung JH, Nava-Ocampo AA, Koren G. Fetal and neonatal outcomes in women reporting ingestion of licorice (Glycyrrhiza

uralensis) during pregnancy. Planta Med. 2013;79:97–101. doi: 10.1055/s-0033-1352464. [PubMed] [CrossRef] [Google Scholar] [Ref list]

75. Kao T-C, Wu C-H, Yen G-C. Bioactivity and potential health benefits of licorice. J Agric Food Chem. 2014;62:542–553. doi: 10.1021/jf404939f. [PubMed] [CrossRef] [Google Scholar] [Ref list]

76. Russo S, Mastropasqua M, Mosetti MA, Persegani C, Paggi A. Low doses of liquorice can induce hypertension encephalopathy. Am J Nephrol. 2000;20:145–148. doi: 10.1159/000013572. [PubMed] [CrossRef] [Google Scholar] [Ref list]

77. Zani F, Cuzzoni MT, Daglia M, Benvenuti S, Vampa G, Mazza P. Inhibition of mutagenicity in Salmonella typhimurium by Glycyrrhiza glabra extract, glycyrrhizinic acid, 18alpha- and 18beta-glycyrrhetinic acids. Planta Med. 1993;59:502–507. doi: 10.1055/s-2006-959748. [PubMed] [CrossRef] [Google Scholar] [Ref list]

78. Asl NM, Hosseinzadeh H. Review of pharmacological effects of Glycyrrhiza sp, and its bioactive compounds. Phytother Res. 2008;22(6):709–724. doi: 10.1002/ptr.2362. [PMC free article] [PubMed] [CrossRef] [Google Scholar] [Ref list]

I want morebooks!

Buy your books fast and straightforward online - at one of world's fastest growing online book stores! Environmentally sound due to Print-on-Demand technologies.

Buy your books online at
www.morebooks.shop

Kaufen Sie Ihre Bücher schnell und unkompliziert online – auf einer der am schnellsten wachsenden Buchhandelsplattformen weltweit! Dank Print-On-Demand umwelt- und ressourcenschonend produziert.

Bücher schneller online kaufen
www.morebooks.shop

KS OmniScriptum Publishing
Brivibas gatve 197
LV-1039 Riga, Latvia
Telefax: +371 686 204 55

info@omniscriptum.com
www.omniscriptum.com

Printed by Books on Demand GmbH, Norderstedt / Germany